REPORT

T0146343

Officer Classification and the Future of Diversity Among Senior Military Leaders

A Case Study of the Army ROTC

Nelson Lim, Jefferson P. Marquis, Kimberly Curry Hall, David Schulker, Xiaohui Zhuo

Prepared for the Office of the Secretary of Defense

RAND NATIONAL SECURITY RESEARCH DIVISION

The research described in this report was prepared for the Office of the Secretary of Defense (OSD). This research was conducted under the auspices of the Forces and Resources Policy Center of the RAND National Defense Research Institute, a federally funded research and development center sponsored by the OSD, the Joint Staff, the Unified Combatant Commands, the Department of the Navy, the Marine Corps, the defense agencies, and the defense Intelligence Community.

Library of Congress Cataloging-in-Publication Data

Officer classification and the future of diversity among senior military leaders : case study of the Army ROTC / Nelson Lim ... [et al.].
    p. cm.
    Includes bibliographical references.
    ISBN 978-0-8330-4802-8 (pbk. : alk. paper)
    1. United States. Army. Reserve Officers' Training Corps—Case studies. 2. United States—Armed Forces—Minorities. 3. United States—Armed Forces—Officers. 4. United States—Armed Forces—Personnel management. 5. Diversity in the workplace—United States. I. Lim, Nelson.

U428.5.O37 2009
355.3'3108900973—dc22

                                      2009039537

Published 2009 by the RAND Corporation
1776 Main Street, P.O. Box 2138, Santa Monica, CA 90407-2138
1200 South Hayes Street, Arlington, VA 22202-5050
4570 Fifth Avenue, Suite 600, Pittsburgh, PA 15213-2665
RAND URL: http://www.rand.org/
To order RAND documents or to obtain additional information, contact
Distribution Services: Telephone: (310) 451-7002;
Fax: (310) 451-6915; Email: order@rand.org

Preface

U.S. Department of Defense (DoD) officials have expressed concern about the relative scarcity of minorities in the senior leadership of the military. This report examines the proposition that this disparity is partly attributable to the process by which officers choose their career fields. In sum, officers who reach the top ranks of the military tend to come from career fields that are disproportionately occupied by whites, and the relative lack of minorities in these fields has a significant impact on the diversity of the senior leadership. While not offering a definitive conclusion, this report suggests that there is a relationship between career field selection, racial/ethnic status, and membership in the senior officer corps. Moreover, this connection appears to stem, at least in part, from racial/ethnic differences in the occupational preferences of officer cadets.

This report is intended for DoD policymakers interested in personnel diversity in the armed forces and is the final product of an exploratory study funded by RAND National Defense Research Institute (NDRI) research support funds. The study was conducted under the auspices of NDRI's Forces and Resources Policy Center. NDRI is a federally funded research and development center sponsored by the Office of the Secretary of Defense, the Joint Staff, the Unified Combatant Commands, the Department of the Navy, the Marine Corps, the defense agencies, and the defense Intelligence Community. Nelson Lim served as the principal investigator. Comments are welcome and may be addressed to Nelson_Lim@rand.org. For more information on RAND's Forces and Resources Policy Center, contact the Director, James Hosek. He can be reached by email at James_Hosek@rand.org; by phone at 310-393-0411, extension 7183; or by mail at the RAND Corporation, 1776 Main Street, Santa Monica, California 90407-2138. More information about RAND is available at www.rand.org.

Contents

Figures

Tables

Summary

Throughout recent history, the U.S. military has served as a model for racial integration and has seen diversity flourish in its organization. Still, while the enlisted ranks of the U.S. military exhibit a high level of demographic diversity, the leadership of the military has remained demographically homogenous.

This report summarizes findings from an exploratory study of a potential barrier to improving demographic diversity in the senior officer ranks. We started with an observation that officers with combat-related career backgrounds, such as the Combat Arms branches of the U.S. Army, tend to populate the top levels of the Army. In 2006, for instance, 80 percent of Army generals were in Combat Arms branches. We also observed that minority officers are disproportionately absent in these key career fields.

This report touches on the career field assignment processes for all services and commission sources. We found that each military service (Army, Air Force, Navy, and Marine Corps) and commission source (Reserve Officers' Training Corps [ROTC], service academy, and officer training school/officer candidate school) has a distinct career field assignment process. However, because we had limited resources, we concentrated on the Army ROTC process as a detailed case study. As a case study, its results are not fully generalizable to other services or even to other commission sources within the Army. Our primary aims are to highlight the importance of the issue and to motivate the U.S. military to conduct a comprehensive study of the issue.

The Army ROTC Career Field Assignment Process

In this report, we look in detail at the Army ROTC classification process. As the first step in this process, the Army obtains career field preferences from ROTC cadets. Each cadet ranks his or her top choices of career fields, without restrictions on academic major or a qualifying test. The Army then combines these preference rankings with the cadet Order of Merit List (OML). The OML ranks cadets according to academic achievement, leadership, and physical fitness. The top 10 percent of the OML automatically receive their top preference. For the remaining 90 percent of cadets, the Army moves down the list and places each cadet in his or her first-choice career field until that field has reached its quota for the year. If a cadet's first choice is full, the Army assigns the second choice, then the third choice, and so forth, as the Army continues down the OML and career fields continue to fill up.

A few complications occur throughout the process. While the career assignment process for the top half of the OML is rigid, there is more flexibility in the lower half. In an attempt to distribute quality across career fields, the Army employs the *65 percent rule*. This rule allows for no more than 65 percent of any one branch's entry-level requirements to be filled from the top half of the OML. Thus, cadets in the top half of the OML whose first choices are 65 percent full will receive the next feasible choice, while the remaining 35 percent of those career fields will go to cadets in the bottom half. This rule allows some lower-quality cadets to enter popular career fields.

In addition, cadets have the option to volunteer for the Branch for Service program, which extends their active duty service obligation (ADSO) by three years. After the first half of the OML list has been assigned, the Army gives those who volunteer for this program preference over those who do not. Finally, the active component Department of the Army Selection and Branching Board plays a role in assigning lower-quality cadets.

Consequences of Cadet Career Preference and Cadet Quality

As described above, there are two main factors in a cadet's career field assignment in the Army ROTC program: cadet career preferences, which are obtained directly, and cadet quality as measured by the OML. Our quantitative analysis examined how these factors affect minority representation across career fields. We used 2007 Army ROTC branching board results and concentrated on male cadets.[1]

Our analysis showed that career field preferences do differ across racial/ethnic lines. For instance, African American cadets tend to prefer Combat Service Support branches whereas white cadets most often opt for Combat Arms branches.

Next, we determined that minorities tend to rank lower on the OML. In general, whites earned higher Order of Merit scores (OMS) (which determine a cadet's rank on the OML) than did minority cadets. While more than 10 percent of white cadets ranked in the top 10 percent of the OML in 2007, fewer than 3 percent of African Americans ranked as high on the list.

Finally, we examined whether cadets of different racial/ethnic groups were receiving their top career field preferences. When we compared cadets' ranked preferences with the career fields to which they were actually assigned, we found that, regardless of race/ethnicity or rank on the OML, most cadets received one of their top career field choices. Therefore, we concluded that racial and ethnic differences in career preferences largely explain the low number of minorities in the Combat Arms branches.

However, it is important to note that, even though we have shown that minorities tend to indicate preferences for different career fields than do whites, we do not know the reasons behind these preferences. For example, we do not know whether minority cadets actually like Combat Service Support branches or whether they adjust their branch choice according to how competitive they are. To examine this possibility, we looked at how the percentage of cadets who picked a Combat Arms branch varied with OMS. Hispanic and other race/ethnicity cadets appeared more likely to choose a Combat Arms branch as their competitiveness increased, but white and African American cadets showed no clear relationship between

---

[1]   Our analysis concentrated on male cadets, for the following reasons: (1) Most Combat Arms positions are not open to female officers, so their career field selection process is not directly comparable. (2) Only 498 female cadets were observed in our data, making up less than 17 percent of the 2007 Army ROTC cohort. (3) Female cadets are not distributed evenly across racial/ethnic groups.

OMS quartile and propensity to opt for a Combat Arms branch as their top choice. This result suggests that there may be true differences in preferences across racial/ethnic groups that are not explained by individual competitiveness, although the limited statistical power of our small sample did not allow us to definitively address this question.

Policy Discussion and Recommendations

Although our analysis is merely a single case study, the results suggest a general policy recommendation for all services.

We recommend that DoD conduct a comprehensive study of its classification systems within all services and commission sources because we have shown that a lack of minorities in key career fields can be one of the major barriers in improving diversity among the top military leaders.

In addition, our findings imply that the Army has three options to improve the level of racial/ethnic diversity in the top officer ranks:

1. Promote more officers from Combat Support and Combat Service Support career fields.
2. Disproportionately promote minorities in the Combat Arms career fields.
3. Increase the number of minorities in Combat Arms.

The first option requires a fundamental change in the Army culture. The second option explicitly inserts race/ethnicity into the promotion process. The third option seems the most feasible, but the Army would need to adjust the incentives for choosing Combat Arms branches in a way that appeals to minorities.

Currently, the order of merit system does not appear to prevent minorities from entering their preferred career fields. If further research highlighted a policy option that could shift the preferences of minorities toward Combat Arms branches, such a policy might bring the problem of their lower OML rankings into play. If minorities began to prefer Combat Arms at rates similar to those of white cadets, they may no longer receive their top preferences, since white cadets (whose OML rankings tend to be higher) also prefer Combat Arms. In this case, the Army would need to modify the order of merit system to get more minorities into Combat Arms branches.

Although it is clear that certain aspects of support branches are more appealing to minority cadets than to white cadets, unless we can identify which incentives minorities respond to, we will be unable to increase the number of minorities who prefer the Combat Arms branches. And, as we pointed out above, we still need to understand the reasons behind minority cadets' career choices. A greater understanding of these reasons will allow us to craft more effective policies to increase the number of minorities in key career fields. Therefore, we strongly recommend that the Army conduct a comprehensive study of its classification system, including an analysis of cadets' career preferences.

The military cannot create general officers overnight. Many factors in addition to career field choice affect whether or not an officer can successfully reach the highest levels of leadership. There are no guarantees that simply shifting minorities into preferred career fields will produce minority generals. Still, this report demonstrates that career field selection is one mechanism within the military that influences the long-run diversity of the senior leadership, and therefore, one that demands examination from policymakers.

Acknowledgments

We thank all the military services for providing information on their officer classification systems and we are particularly grateful to the Army for providing the data we used in our analysis. We especially thank Col James Campbell of the DoD Office of Diversity Management and Equal Opportunity for his help in the production of this report. We are very grateful to our RAND colleagues Beth Asch and Lynn Scott for reviewing our work and helping us to improve it. We thank Aaron Martin for providing useful information on Army ROTC as well as helping us to proof the report, and we thank Catherine Chao for helping us prepare the document for publication.

Abbreviations

ADSO	active duty service obligation
AFSC	Air Force Specialty Code
APS	academic program score
CPR	cadet performance rank
CPS	cadet performance score
DA	Department of the Army
DCS	Deputy Chief of Staff
DHRB	Defense Human Resources Board
DMDC	Defense Manpower Data Center
DMG	Distinguished Military Graduate
DoD	Department of Defense
DWG	Diversity Working Group
HRC	Human Resources Command
MPS	military program score
NRL	nonrated line
OCS	Officer Candidate School
ODMEO	Office of Diversity Management and Equal Opportunity
OML	Order of Merit List

OMS	Order of Merit score
OPMS	Officer Personnel Management System
OSD	Office of Secretary of Defense
OTS	Officer Training School
PPS	physical program score
ROTC	Reserve Officers' Training Corps
SES	Senior Executive Service
TBS	The (Officer) Basic School
USAFA	United States Air Force Academy
USMA	United States Military Academy
USNA	United States Navel Academy

CHAPTER ONE

Introduction

Historically, the enlisted ranks of the U.S. military have had a more diverse racial and ethnic population than that of the officer ranks. During the Vietnam era, some Department of Defense (DoD) critics attributed the relative paucity of minority officers to a systematic policy of racial discrimination, which contributed to morale problems and heightened racial tensions within the military. Consequently, recent DoD leaders, many of who served as junior officers in the 1960s and 1970s, have recognized the importance of a diverse officer corps. For example, in 2003, 29 former military and civilian leaders of DoD—including several retired four-star generals, chairmen of the Joint Chiefs of Staff, and secretaries of defense—filed an Amicus Curiae brief, urging the Supreme Court, in the case of *Grutter v. Bollinger,* to uphold University of Michigan law school's affirmative action plan (Groner, 2003). In their brief, these leaders maintained that a highly qualified, diverse military leadership was essential to U.S. national security. To fulfill its mission, they asserted, the military "must be selective in admissions for training and education for the officer corps, *and* it must train and educate a highly qualified, racially diverse officer corps in a racially diverse setting" (Mason, 1998, p. 3). These leaders supported this claim by citing the negative impact that past perceptions of discrimination have had on troop morale.

DoD has made great strides in improving officer corps diversity during the past 40 years. Between 1967 and 1991, the Pentagon almost quadrupled the minority representation in the ranks of its newly commissioned officers, and the proportion of female officers increased ninefold (Hosek et al., 2001, p. xiii). From 1986 to 2006, minority officer representation increased nearly 5 percent; at the highest levels (O-7 and above), minority representation increased 9 percent (DMDC, 2006).

Still, the relative scarcity of minority military leaders remains an issue, probably because the U.S. government usually bases its arguments for demographic representation on issues of *access and legitimacy* (Kraus and Riche, 2006). The government has historically argued that representation of minority groups is important because it demonstrates that the public policy realm is open to and representative of all people. In addition, DoD is concerned that no particular group should bear the costs and sacrifices of military service unequally (Kraus and Riche, 2006). Therefore, demographic comparisons at each level of the military hierarchy are often relevant to discussions of diversity in the military.

Although the military has seen significant gains in racial and ethnic representation throughout the military, the most senior levels still do not fully reflect these gains. As an illustration, Figure 1.1 shows a recent picture of how the racial and ethnic distribution of officers

Figure 1.1
Racial/Ethnic Distribution in the Enlisted, Officer, and Senior Officer Ranks in 2006

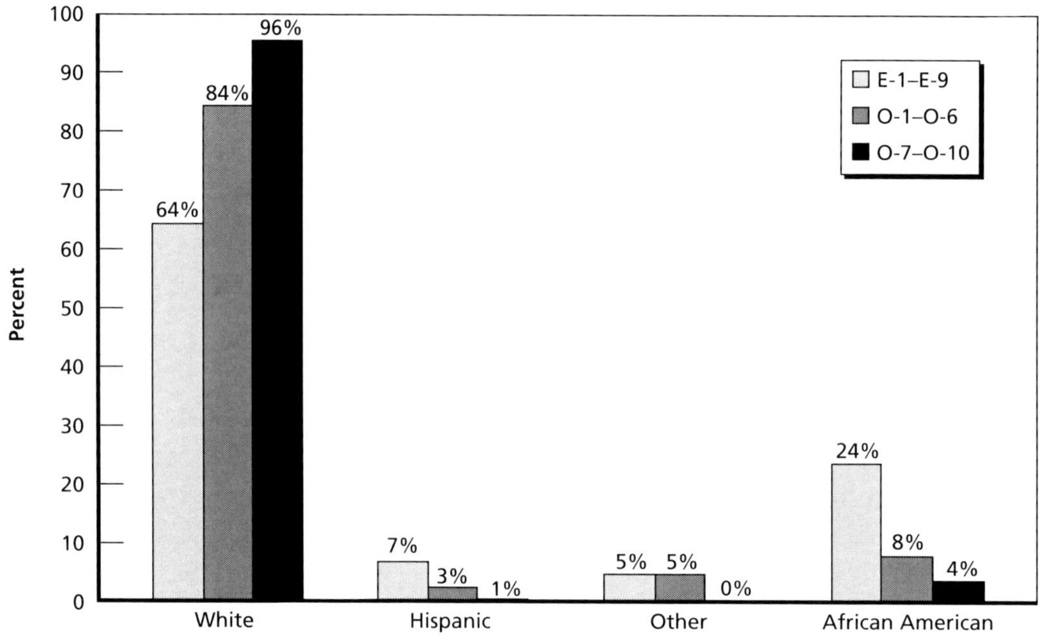

SOURCE: DMDC PERSTEMPO Database.
RAND TR731-1.1

compares to the enlisted force.[1] In 2006, 31 percent of the enlisted ranks of the military were African American or Hispanic (24 percent African American, 7 percent Hispanic). However, only about 16 percent of all officers were African American or Hispanic (12 percent African American, 4 percent Hispanic), and only 5 percent of all O-7–O-10s were African American or Hispanic (4 percent African American, 1 percent Hispanic) in 2006.

The literature on military personnel policy points to a range of factors to explain a lack of minorities in the officer corps. Clearly, the lower college graduation rates of minorities compared to those of whites restrict the pool of potential minority officers (Baldwin, 1996b). Minority group members may also have lower promotion rates within the military because their records do not indicate the same level of achievement as their majority counterparts (Baldwin and Rothwell, 1993). Other research suggests that minority officers face greater difficulties in forming peer and mentor relationships (which are vital to success in the military) and that minority officers often must serve in a recruiting capacity (to recruit more minorities), giving them less experience (Hosek et al., 2001).

A few researchers at least tacitly acknowledge the impact of occupational segregation on minority and female promotion rates. Stewart and Firestone (1992) indicate that African American officers are concentrated in the technical/operational job category; most female offi-

---

[1]   While the comparisons in Figure 1.1 are informative about the racial/ethnic makeup of the different tiers of the military, it is important to note that other comparisons might paint a different picture of whether the military has a relative scarcity (or possibly a relative surplus) of minority personnel. For example, we could compare the highest levels of military leadership with corporate executives having similar amounts of responsibility. The purpose of Figure 1.1 is to illustrate the policy problem that motivates this research and not to conduct a complete assessment of the level of under- or overrepresentation of minorities in the military ranks. Furthermore, increasing minority representation could still be a goal for DoD leaders, regardless of the current level of representation compared with relevant benchmarks.

cers, in the medical/dental category. As a result, they state the Army and Navy are considering revising their branching procedures to achieve a more representative distribution of minorities and women. Chestang (2006) claims that DoD's new Officer Personnel Management System (OPMS) XXI will benefit minorities and women by providing more promotion opportunities for those outside the Combat Arms career fields. While research has acknowledged the reality that minorities do not tend to concentrate in highly promoted career fields, none has examined why this is the case by looking at cadet career field choices.

Impetus for Achieving a More Diverse Senior Leadership

In light of the previously highlighted scarcity of minority senior leaders, increasing racial, ethnic, and gender diversity has become a priority outside and inside DoD. Members of Congress have inquired about DoD's efforts on diversity, and other external observers have highlighted DoD's challenges with respect to the retention and promotion of minorities and women (Lubold, 2006; Hosek et al., 2001; Baldwin, 1996a, 1996b; Meek, 2007). In May 2005, then Secretary of Defense Donald Rumsfeld issued a directive to "put much more energy into achieving diversity at senior levels of services" (Diversity Working Group, 2005). In 2007, the Office of the Secretary of Defense (OSD) and the Office of Diversity Management and Equal Opportunity (ODMEO), with assistance from RAND, brought together diversity experts from academia and the public and private sectors to meet with DoD representatives for two days of discussion and inquiry on diversity issues. At this meeting, participants pointed out that as an organization that promotes from within, DoD's top leadership is dependent upon the pipeline of junior officers. Looking at this pipeline, they found no prospect for an increase in the representation of minorities or women in the higher ranks (flag officers and Senior Executive Service [SES] members) for the next decade. In other words, labor force trends will not cause an increase in minority senior leaders without some kind of policy intervention, and the divergence between the general population and those in charge of the military is likely to worsen if nothing is done. In order to avoid the negative consequences that accompany a lack of equal opportunity (real or perceived), participants recommended identifying barriers to minorities' attainment of top leadership positions.

There was a consensus among the experts at the ODMEO meeting that military career fields (or occupations) that are combat-related (or operations-related) tend to promote to the highest levels of the military. For example, 80 percent of Army generals come from the Combat Arms branches (see Figure 1.2[2]).

In general, there has been a tendency within the military for a particular genre of career fields (usually the career fields that are most closely identified with the overall mission) to form a professional elite. Members of that elite set the policies and procedures and determine the criteria for promotion, and wielding of this power in any organization can tend to perpetuate the dominance of the elite group (Mosher, 1982). In addition, promotions in the military are competitive and merit-based and also involve increasing levels of strategic responsibility, so combat experience will naturally improve an individual's promotion chances.

---

[2]   Figure 1.2 shows the current branch distribution for Army generals, whereas this report focuses on the entry branch for Army officers. Generals who are currently in Combat Arms branches could have initially entered a different branch category, so this computation is an estimate of the entry branch distribution (which is potentially biased).

Figure 1.2
Branch Distribution of Army Generals (O-7 and Above) in 2006

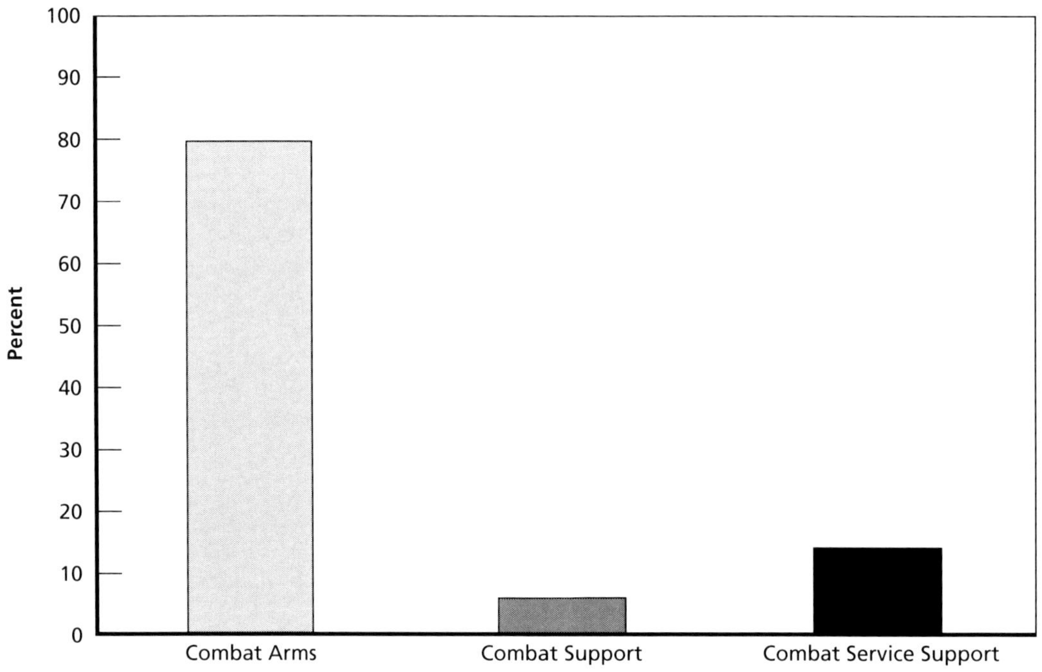

RAND *TR731-1.2*

Participants at the diversity meeting noted that, for the most part, women and minorities are not in these combat-related career fields that tend to promote to senior leadership. For example, Figure 1.3 shows the branch distribution of new Army officers (O-1s) by race/ethnicity. While a majority of whites (56 percent) and a plurality of Hispanics (49 percent) are found in the Combat Arms branches, the highest percentage of African Americans (40 percent) are located in Combat Service Support occupations; only 34 percent are in Combat Arms.

The prevalence of whites in Army Combat Arms branches is even more pronounced in the case of experienced officers (see Figure 1.4). At the O-6 level, whites are the only racial or ethnic group with a plurality serving in Combat Arms branches (47 percent). Most of the members of other racial/ethnic groups serve in Combat Service Support branches (69 percent of African Americans, 52 percent of Hispanics, and 67 percent of other minority groups). Only 22 percent of African Americans, 28 percent of Hispanics, and 33 percent of other minority groups are in Combat Arms.

Report Focus

As stated above, a correlation exists between the occupational specialty to which an officer candidate is assigned and how far he or she progresses up the military career ladder—with the top levels of military leadership mostly stemming from a short list of "favored" career fields. To better understand the relationship between career fields and the racial/ethnic makeup of the

Figure 1.3
Branch Distribution of New Army Officers (O-1) in 2006

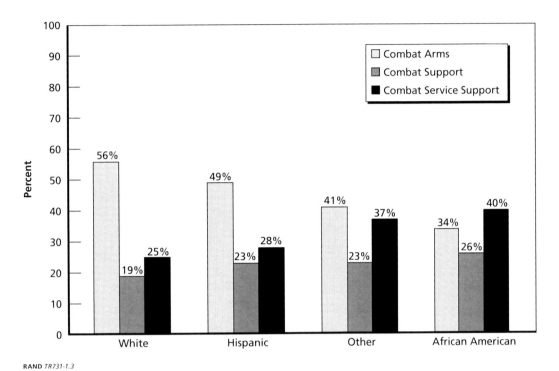

RAND *TR731-1.3*

Figure 1.4
Branch Distribution of Experienced Army Officers (O-6) in 2006

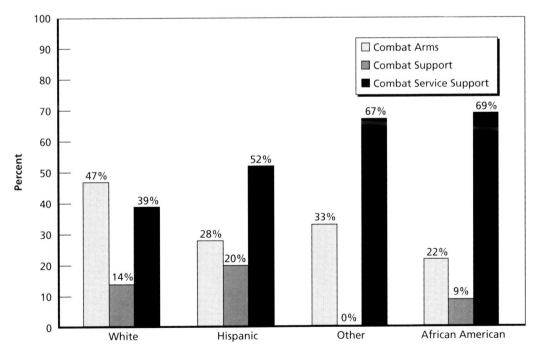

RAND *TR731-1.4*

senior military leadership, this report examines the occupational selection process for officers, the involvement of minorities in this process, and process outcomes with respect to diversity.

We focus primarily on the Army Reserve Officer Training Corps (ROTC) career assignment process, for several reasons. First, the Army's classification system has few entry requirements and barriers. In other words, most Army ROTC cadets can choose from all the branches without having to meet specialized requirements. This simplicity allows for the analysis of an entire cohort at once, without the need to break the group down into a multitude of separate special cases. Second, the Army ROTC branch assignment process provides a transparent example of how individual preferences and structural rules can interact in a way that produces racially/ethnically divergent outcomes. Finally, the Army ROTC selection data were the most comprehensive data available for our analysis.

It is also important to note the major limitation of a single case study approach: these results do not necessarily generalize to other commissioning sources or other services. For example, cadets at the United States Military Academy (West Point) and Army Officer Training School (OTS) may have different motivations behind their career choices and may face different rules and constraints for their branch assignment process. In addition, the incentives that draw minorities into particular Army branches could be completely different from the incentives that draw minorities into similar Navy, Marine Corps, and Air Force career fields.

Therefore, although an analysis of Army ROTC branch assignment provides a clear example of how to think about the racial/ethnic differences in career field assignment, these specific findings are not robust enough to inform Army- or DoD-wide policy.

Organization of the Report

The remainder of the report is organized as follows. Chapter Two draws on social science literature to provide a theoretical background on the individual determinants of career field decisions. Chapter Three investigates the career field assignment process in order to understand how we might alter the structure to change outcomes for particular groups. We broadly examine the four services' classification processes, discuss the major differences between the services' and commission sources' processes, and analyze the Army ROTC branching in detail. Going beyond process, Chapter Four explores the underlying reasons why minorities are scarce in career fields associated with the top levels of military leadership. Through a quantitative analysis of the Army ROTC branching process, we examine the branch assignments of different demographic groups and attempts to determine whether these assignments reflect cadet preferences, qualifications, and/or some other factor. Finally, Chapter Five presents recommendations for addressing the lack of diversity in the military's senior leadership.

CHAPTER TWO

Theoretical Determinants of Individual Career Field Decisions

The social science literature on occupational selection provides a useful starting point for thinking about factors that influence the racial/ethnic distribution of officers across career fields. It will first help to frame the military officer occupational decision in the context of economic and sociological theory.

Sociologists have been interested in occupational structures since the founding of the discipline. For instance, Max Weber, a German scholar considered by many to be one of the founders of sociology, distinguished between two types of occupational structures: *free* and *unfree*. He writes the following:

> The unfree organization of occupations exists in cases where there is compulsory assignment of functions within the organization. . . . The free type of distribution arises from the successful offer of occupational services on the labor market or successful application for free "positions" (Weber, 1947).

By this definition, some elements of the military occupational structure are unfree, because service requirements can force some individuals into occupations that they would not otherwise have chosen. But, in the era of all volunteer force, individuals in the U.S military still have some influence over their career development, and in this sense the system is at least partly free.

In addition to the different types of occupational structures, sociologists distinguish between two types of competition: wage competition and vacancy competition. *Wage competition* describes labor markets in which employers offer a wage that is equal to the (marginal) productivity of the worker and individuals compete to receive that wage. *Vacancy competition* refers to situations in which individuals have access to jobs only when others leave, and employers fill job vacancies based on an employee's relative position in the organizational hierarchy (also known as the "job queue") (Kalleberg and Sørensen, 1979; Sørensen and Kalleberg, 1981). Military labor markets contain some elements of each type of competition. Individuals do compete for wages in the sense that some jobs have bonuses for being additionally productive (though this varies by service), but new jobs tend to be vacancy driven and filled from a job queue within the organization.

Within the broader classification structure, occupational decisions still take place at the individual level. Most of the social science theory on individual occupational choice boils down to a few basic precepts of human behavior. An individual choosing an occupation will have a variety of options available. The options will depend on numerous factors, including the person's qualifications and desires and the availability of work. From the list of career field options, an individual then compares the dimensions of each option that are most important. For instance, some career fields could provide useful skills that may pay off in the future. Other career fields may include a high level of social prestige. Some career fields come with more risk (injury, failure, etc.), so people may prefer the safer options. Ultimately, people weigh the different options according to their own criteria and choose the best one. Most social science theory and research stems from this basic framework.

For example, Blau et al. (1956, p. 533) note that an individual's choice is based on his "valuation of the rewards offered by different alternatives" combined with his "appraisal of his chances of being able to realize each of the alternatives." In a military context, this would mean that cadets form their preferences, which they always submit in some capacity to the military prior to an assignment, by balancing their actual preferences with their individual estimate of how likely they are to actually get each preference.

Economists view decisions in a similar fashion. Economic theory supposes that individuals will choose the career field that maximizes their utility. *Utility* is a measure of individual satisfaction that depends on all the characteristics of each option that an individual values (or wishes to avoid). For example, individuals may receive additional utility from a higher bonus but might lose utility from rigorous job requirements.

Previous RAND work has viewed enlistment decisions in this fashion (see Kilburn and Klerman, 1999; Hosek and Peterson, 1990). This work assumes that individuals choose the career path that maximizes their utility, weighing enlistment against additional schooling and/or work in the civilian labor force. Therefore, the relative attractiveness of civilian opportunities (in addition to different job options within the military) will necessarily affect an individual's career decisions. Furthermore, an individual's decision or plans to remain in the military must interact with his/her career field decisions.

Many variables could affect an individual's utility for different occupations. Although class origin and educational attainment limit some individuals to certain career fields, social scientists also recognize the importance of additional factors (some interrelated) that affect individuals' occupational decisions. One such factor, which many studies on racial/ethnic differences in military career field selection cite, is civilian job transferability. Microeconomic theory refers to this aspect of the decision as an investment in *human capital*. Human capital investment includes all "activities that influence future real income through the imbedding of resources in people" (Becker, 1962). Some career fields provide training, experience, and skills that individuals can take with them after they finish serving in the military. Therefore, people who plan to enter the civilian labor force in the near future may prefer occupations that include transferable investments in their human capital stock. Furthermore, some studies offer the hypothesis that minorities, in particular, place a high value on transferable human capital investment. If this hypothesis were true, it would explain why minorities tend to sort themselves into the more-transferable noncombat career fields (Harrell et al., 1999; Hosek et al., 2001). However, Hosek et al. note that African Americans are less likely to leave the military, which sheds some doubt on the hypothesis that they join to acquire skills for the civilian job market. Human capital theory predicts just the opposite: Acquiring more transferable human capital should *increase* the amount of job turnover (Becker, 1962).

However, very few people make completely isolated decisions. The body of social science research recognizes that individuals make decisions in a social context. Another factor, then, that could potentially influence an individual's occupational decision is social approval. If an individual's social group revered some occupations more than others, the individual would likely prefer more social prestige. Nobel Laureate economist George Akerlof summarized this phenomenon when he wrote, "While my network of friends and relatives are not affected in the least by my choice between apples and oranges, they will be affected by my educational aspirations. . . . As a consequence, the impact of my choices on my interactions with other members of my social network may be the primary determinant of my decision" (1997). Moreover, social approval could easily vary for different racial/ethnic groups, which would explain the observed differences in occupational choice. Mani and Mullin (2004) expand upon this idea, adding that the amount of social approval for a career field may depend on how aware the community is of the skill that the career field requires. Thus, individuals in a community with a history of combat service would benefit from additional social approval by choosing combat career fields. Harrell et al. (1999) note the following anecdotal examples that illustrate this phenomenon vividly:

> In contrast to a majority [white] enlisted discussant who was asked, "Are you going to be a SEAL?" when he mentioned that he was joining the Navy, one minority officer said, "Kids in the inner city schools ask, 'why are you going to swab decks?'" Another discussant said: "I go home and people don't know anything about it. I talk to people in my church—Italians and whites, they know all about it, but blacks don't."

Social approval could be an important factor in occupational decisions, and the level of social approval is likely to differ in degree depending on social group and background, which almost certainly vary by race/ethnicity.

One way in which social perceptions could vary with race/ethnicity has to do with information. The information people have about an occupation is an important factor in their occupational choice (Blau et al., 1956), and people get their information from those around them—parents, community elders, teachers, coaches, or peers (Sewell, Haller, and Portes, 1969). People also gain information by reading books or surfing the Internet. Thus, it is plausible that different occupational information circulates among individuals in different racial/ethnic groups. With military classification, information about previous conflicts that circulates within social networks may have an effect on the occupational preferences of recent minority cadets. For example, during the Vietnam War, African Americans often supplied a disproportionate number of frontline combat troops, where they were in greater danger of suffering casualties than were their white counterparts.[1] Today, a minority cadet who hears a parent or relative express concerns related to this legacy may think twice about choosing a Combat Arms specialty. Survey data show some evidence of this perception: The 1999 Youth Attitude Tracking Study notes that "black youth are more likely to mention threat to life or to say killing is against their beliefs" (Wilson et al., 2000).

There is some empirical evidence that social networks affect individual decisions. In a pertinent example, Martin Kilduff (1992) found that people tend to make career decisions that are similar to those of their friends; the degree of similarity may depend on how the individual relates to others. Although researchers have long known that individual choices are correlated with the individual's social network, it is extremely difficult to tell if the social network is the cause of the choice. The same effect could just as easily result from some other factor that affects everyone in the network or simply from the fact that individuals tend to choose people similar to themselves for friends. Still, some research has found credible evidence of network

---

[1]   In 1968, African Americans made up about 12 percent of the Army and Marine Corps forces, yet they frequently contributed half of the men in frontline combat units, especially in rifle squads and fire teams (Coffey, 1998; see also Rostker, 2006, pp. 320–328).

effects on individual behaviors, such as welfare receipt (Bertrand, Luttmer, and Mullainathan, 2000), education and wages (Borjas, 1995), and labor market participation (Aguilera, 2002).

Based on a synthesis of the social science literature, Figure 2.1 provides a conceptual view of the factors affecting the distribution of officers across career fields within the U.S. military. The figure recognizes that socioeconomic background (including race/ethnicity) plays a seminal role in occupational decisions in the military as it does in other parts of society (Kilburn and Klerman, 1999; Harell et al., 1999). However, individual job selections depend on educational factors (such as academic major) as well as personal career goals and expectations and advice from friends, relatives, and mentors. An individual combines all this information with the appeal of the incentives that Army policy attaches to each career field and chooses the career field that offers the highest expected payoff. Additionally, the military's distinctive (and varied) job classification system has the most immediate impact on officer career assignments by melding the cadets' stated career field preferences and their military qualifications with particular military requirements (e.g., for longer active duty service obligation [ADSO] or a disproportionate number of junior Combat Arms officers).

Furthermore, the only variables in Figure 2.1 that are truly outside the realm of policy are those that define an individual's background. Not only does military personnel policy explicitly determine each step of the job classification process, it also implicitly affects some of the intervening variables. Therefore, policy options could affect either the mechanics of the assignment process or the characteristics of each career field that determine an individual's decision prior to branch assignment. This report provides an example of which type of policy may be most appropriate in light of patterns in assignment data.

Figure 2.1
Overview of Factors Influencing the Distribution of Officers Across Career Fields

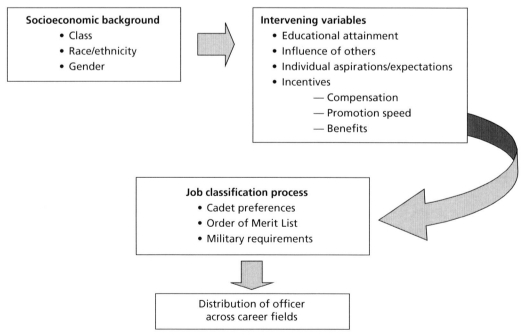

CHAPTER THREE

Career Field Assignment Process: Rules and Practices

As previously shown, a disproportionate number of top-ranking officers are in just a few occupational specialties. Therefore, if we want to know why some cadets are more likely than others to become general or flag officers, it is necessary to understand the process by which cadets are assigned to career fields. In particular, an analysis of the branching process should reveal career assignment variables that will have an eventual effect on the percentage of minority officers who reach the highest levels of the military.

Description of Branching Processes

Each military service (Army, Navy, Air Force, Marine Corps) and commission source (ROTC, service academy, and officer training school/officer candidate school) has a distinct career field assignment (or classification) process. In general, these processes involve cadets providing career preferences and the services ranking cadets according to several qualitative and quantitative factors (see Figure 3.1). For the most part, the list of rankings determines the order in which cadets are assigned to career fields. Starting at the top of the ranking list, the classification systems assign cadets their first preference until career field quotas are met, whereupon they shift to cadets' second preferences, and so on. In other words, the further down the list a cadet is, the less chance he or she has of being assigned to his or her preferred career field. This process of matching cadets' preferences to career fields is achieved through computer models, review boards, or a combination of the two. Service practices that aim to distribute quality cadets across all career fields, such as limiting the percentage of cadets from the top sector of

Figure 3.1
**The Classification Process Sorts and Matches Cadets' Career Preferences and the Services'
Requirement for Factors Associated with Quality**

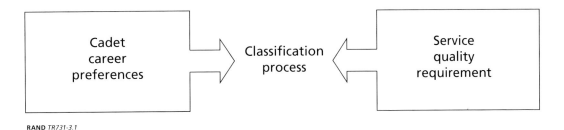

RAND *TR731-3.1*

the ranked list in each occupation, counter the effect of class rankings to some extent. Race/ethnicity is not intended to be a factor in any of the services' classification processes.

The openness of the services' career field assignment processes varies significantly. While the Air Force and Navy restrict career field options for officer candidates, Army cadets are permitted to apply for admission to most career fields.[1] For its part, the Air Force distinguishes between nonrated line (NRL) officers and rated line officers,[2] with each group having its own classification process. Although the NRL is the more accessible of the two lines, it still requires specific academic majors, coursework, or foreign language skills for officers to qualify for many Air Force Specialty Codes (AFSCs).[3] Similarly, Navy midshipmen can apply for admission only to those career field communities for which they have been deemed eligible through special qualifications and exams.

Classification processes also differ in their use of review boards. The Army ROTC Selection Board mainly reviews assignments of cadets at the bottom of the ranked list produced by an automated process. However, review boards play a more prominent role in the service academies, particularly in the Air Force and Navy. Rather than having an algorithm largely determine career field placement, the Air Force Academy and the Naval Academy have review boards that score and rank the candidates, thus leading to greater flexibility (although arguably less objectivity) in the selection processes.

The Army Branching Process

As explained in the introduction, this report focuses on the Army ROTC career field assignment process (see Figure 3.2). Descriptions of the classification (or branching) processes of other services and commissioning sources can be found in Appendix A.

Cadet Preferences and Order of Merit Ranking

In the fall of each year, all Army ROTC cadets expecting to graduate provide career field preferences without regard to academic major or the results of special qualifying tests. The Order of Merit List (OML) ranks cadets according to academic achievement (40 percent), leadership (45 percent), and physical fitness (15 percent).[4] The top 10 percent of cadets on the OML automatically receive their first career field choice, and placement on the OML generally determines whether the remaining cadets will obtain one of their top choices. If minorities tend to rank lower on the OML, they should be less likely to receive their job preferences. Thus, improving the OML ranking of minorities could be important to increasing their promotion opportunities.

The Branch for Service Program

At the time preferences are collected, Army ROTC cadets can volunteer for the Branch for Service program, which extends their ADSO by three years. After the first half of each

---

[1]   Restrictions on women in Army Combat Arms branches are discussed in Chapter Four.

[2]   Rated line officers are pilots, navigators, and air battle managers.

[3]   AFSCs specify the types of billets an officer may fill.

[4]   These OML components are further broken down in the appendix.

Figure 3.2
Flowchart of Army ROTC Branch Allocation Methodology

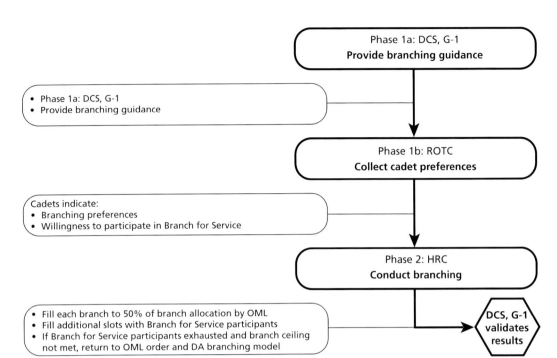

SOURCE: Denning, 2006.
RAND *TR731-3.2*

branch's allotments is filled in accordance with the OML, remaining Branch for Service cadets are classified, bypassing higher-ranked cadets who do not opt for the additional ADSO. This program is an excellent opportunity for those who rank lower on the OML to increase their chances of obtaining their preferred career field.

The 65 Percent Rule and the Role of Boards

The Army ROTC selection process allows no more than 65 percent of any one branch's entry-level requirements to be filled from the first half of the OML, leaving 35 percent of each career field's slots available to the lower half of the OML. This practice distributes at least some high-quality junior officers across all career fields and permits some lower-ranking candidates to enter popular career fields.

Selection boards play a greater role in the classification of lower-ranking cadets. After classifying the top half of the OML and assigning branches to ADSO volunteers, the Department of the Army (DA) branching model matches the remaining cadets to career fields. The active component DA Selection and Branching Board reviews these allocations and makes the final decision about branching assignments. Although the top of the OML list is branched according to fairly rigid criteria, the board allows for more flexibility in assigning jobs to those in the bottom half of the OML. To the extent possible, the board ensures that lower-ranking candidates receive at least one of their career field preferences. If minorities tend to fall lower on the OML, then review boards could be important to providing some minority officers the opportunity to be admitted into high-status career fields. However, these minority officers would have to express a preference for one or more of those fields.

Preferences Versus Ranking: The Role of Policy

The branching process attempts to maximize the Army's effectiveness by balancing individual cadet desires with the Army's personnel constraints. In this framework, it is tempting to treat individual preferences for specific career fields as exogenous (i.e., beyond the realm of policy intervention). This view is incorrect, however, because individual tastes for certain occupations are only a part of what goes into an individual's career field decision. When individuals choose a career field, they balance their own occupational tastes with a bundle of other characteristics that current Army policy determines. For example, a cadet who chooses a Combat Arms branch not only gets to perform a certain type of job on a daily basis but also receives the benefit of more promotion opportunities—stemming from prior policy decisions. In this sense, cadet preferences are not fixed but can be altered by changing the incentives that go along with each career field. Thus, policy options to alter the racial/ethnic makeup of a specific career field could address either the incentives that determine individual preferences or the mechanics of whether or not cadets actually obtain their preferences. The best policy depends on whether preferences or quality drives the racial/ethnic differences in branch assignment.

Summary

We have shown that there are two main factors in the Army ROTC classification process: cadet preferences and cadet quality ranking. To determine why minorities are not being placed into the top career fields more frequently, we must examine both of these factors in more detail. If minority cadets are self-selecting into Combat Support and Combat Support Service occupations and their selections are largely being honored by the Army's classification process, then cadet preference is chiefly responsible for the low number of minority officers in Combat Arms, from which most of the Army's generals emerge. Furthermore, if minorities are opting out of Combat Arms, we must ask which incentives are chiefly responsible for funneling them into the Combat Support and Combat Service Support branches. Regardless, if cadet preference is the culprit, then the Army should examine the current policy-driven incentive bundles and consider a strategy for increasing the attractiveness to minorities of the Combat Arms branches as a means of improving minority representation in the general officer corps.

However, if minority cadets are not being admitted to Combat Arms branches because of their relatively low placement on the OML list, then a different course of action is necessary to achieve greater diversity within the senior ranks of the military. This action could include amending the branching process or finding ways to improve minority OML rankings. Finally, a specific combination of cadet preference and OML ranking may be responsible for the disproportionate number of minorities in the support branches and consequent lack of diversity in the senior Army officer ranks. To understand which of the foregoing hypotheses is more likely to be true, Chapter Four analyzes Army personnel data relating to the ROTC officer career field assignment process.

Career Field Assignment Process: Quantitative Analysis

Having documented the U.S. military's officer classification rules and practices, we now examine the results of this process within the context of the Army ROTC. As previously stated, a disproportionate number of top-ranking officers come from a few career fields. Because Combat Arms officers are much more likely to be promoted to the top ranks of the Army, it is important to understand how minority officers are distributed across occupations. This chapter attempts to address two major questions: To which kinds of occupations are minority and majority ROTC cadets predominantly assigned? And, to facilitate improvements in military leadership diversity, what are the reasons behind this distribution?

Racial/Ethnic Disparity Across Career Field Assignments

As we indicated earlier, races/ethnicities are not evenly distributed across the Army branches. Table 4.1 shows that white cadets were concentrated in Combat Arms in 2007, while minority cadets were more likely to be found in Combat Support and Combat Service Support branches. Specifically, 42 percent of African Americans were in Combat Arms, compared to 55 percent of whites and 48 percent of Hispanics. Conversely, almost 37 percent of African Americans were assigned to Combat Service Support fields, while fewer than 25 percent of whites and Hispanics were found in this branch category.

Figure 4.1 displays distributions of race/ethnicity across specific career fields. As shown, in 2007, whites had the highest representation in Infantry, Field Artillery, Corps of Engineers, Armor, and Aviation (all Combat Arms branches). The figure also illustrates the striking difference between the occupational distribution of whites and African Americans. While Hispanics tend to be distributed in a similar pattern to that of whites, African Americans follow a very different pattern. Other racial groups (not represented in Figure 4.1 because of their small numbers) tend to behave similarly to whites with respect to occupational assignment.

As indicated in Chapter Two, there are several potential explanations for why minorities are scarce in career fields associated with the highest levels of the Army officer corps. The next section examines whether divergent preferences or differences in OML rank are chiefly responsible for the lack of minorities in the Combat Arms branches.

Table 4.1
Branch Division of All Male Army Cadets by Race/Ethnicity in 2007 (%)

Race/Ethnicity	Combat Arms	Combat Support	Combat Service Support	Total
White	62.6	13.7	23.7	100
Hispanic	56.9	18.1	25.0	100
Other	50.9	19.6	29.5	100
African American	41.9	21.2	36.9	100
Total	59.8	15.1	25.2	100

NOTES: Other includes Native American, Asian, Asia Pacific Islander, and others.

Figure 4.1
Racial/Ethnic Distribution of Male Army Cadets Across Career Fields in 2007

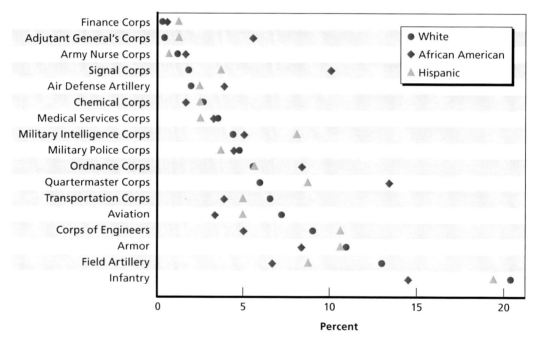

NOTE: Career fields are grouped as follows: Combat Arms: Air Defense Artillery, Armor, Aviation, Corps of Engineers, Field Artillery, Infantry. Combat Support: Chemical Corps, Military Intelligence Corps, Military Policy Corps, Signal Corps. Combat Support Services: Adjutant General's Corps, Army Nurse Corps, Finance Corps, Medical Service Corps, Ordnance Corps, Quartermaster Corps, Transportation Corps.

RAND TR731-4.1

Analytical Results: Army ROTC Case

In an attempt to understand racial/ethnic divisions in classification outcomes in the military, we analyzed 2007 Army ROTC data pertaining to

- racial and ethnic differences in career preferences
- racial and ethnic differences in qualifications
- racial and ethnic differences in obtaining preferred career fields.

Data and Variables

The data for this analysis—and, hence, for all the tables and figures in this chapter—come from the 2007 Army branching board results and include individual information on each Army ROTC cadet's preferences, quality, ranking, final branch assignment, and demographics. The Army G-1 Officer Accessions Policy Branch Division provided the data, and we obtained them on June 1, 2007. Table 4.2 presents some summary statistics for the cadets in our dataset.

These data are especially rich because they include complete information on each part of the branch assignment process. Information on cadet preferences and cadet OML ranking, in particular, allows us to identify the point in the process at which we can see racial/ethnic differences in the outcomes. The data are limited, however, in that they include very little individual background information. The data only include characteristics that factor into the branch assignment process. Thus, they are ideal for a complete description of the assignment process, but they do not permit analysis that seeks to identify the individual mechanisms that ultimately determine the outcomes. In other words, the data allow us to identify whether

Table 4.2
Summary Statistics for 2007 Army ROTC Data

Variable	Mean	Standard Deviation
Male	0.82	0.38
Hispanic	0.07	0.26
Other race/ethnicity	0.08	0.27
African American	0.09	0.29
Order of Merit score	77.87	6.93
Academic grade point average (GPA)	3.09	0.44
Engineering major	0.13	0.34
Humanities major	0.55	0.50
Professional major	0.20	0.40
Science major	0.12	0.32
Attended Senior Military College	0.07	0.25
ADSO volunteer	0.05	0.22
Assigned to Combat Arms	0.50	
Assigned to Combat Support	0.19	
Assigned to Combat Service Support	0.31	

NOTE: Analysis excludes female cadets.

preference or ranking has more of an impact on a cadet's final assignment, but they do not allow us to find out the individual determinants of branch preferences or rankings.

Because female cadets are not distributed evenly across racial/ethnic groups, gender may be one factor affecting the racial/ethnic distribution of cadets across career fields. Table 4.3 shows that a relatively high percentage of African American Army ROTC cadets in 2007 were female (31 percent) compared to a relatively low percentage of white cadets (16 percent). Female representation among Hispanics and other minorities fell in between the other two groups, 21 percent and 19 percent, respectively.

Racial and Ethnic Groups Differ in Their Career Preferences

We found that career field preferences of racial/ethnic groups do differ significantly. Figure 4.2 shows the distribution of cadets' first choice of career field by racial/ethnic group. Figure 4.3 presents these same data by racial/ethnic group and specific branch preference.

The distribution of female cadets within racial/ethnic groups may be a confounding factor in this analysis because women do not face the same conditions as men in the Army's career assignment process. In the Army, women face severe restrictions in entering Combat Arms. Some closures to women are by unit; other closures are by occupation. Although a few artillery units are open to women, aviation is the only Combat Arms branch that is fully open to women.[1] Therefore, we excluded female cadets and only focus on male cadets in the following analysis.

As shown in Figures 4.2 and 4.3, African American cadets' career field preferences are significantly different from those of whites and Hispanics. African American cadets tend to prefer Combat Support and Combat Service Support branches such as Medical Service Corps, Signal Corps, and Adjutant General's Corps. By contrast, white cadets most often prefer Military Intelligence, Infantry, and Aviation. While generally preferring the Combat Support branches, Hispanic cadets make career choices that are closer to those of whites than to African Americans. In sum, Army ROTC minority cadets (especially African Americans) are more likely than their majority counterparts to favor careers in occupational categories that have not been major sources of senior Army leaders.

Table 4.3
Distribution of Cadets by Gender Across Racial/ Ethnic Groups in 2007 (%)

Race/Ethnicity	Male	Female	Total
White	84.2	15.8	100
Hispanic	78.8	21.2	100
Other	81.1	18.9	100
African American	68.9	31.2	100
Total	82.2	17.8	100

---

[1]   Restrictions on women are not consistent across all services. While roughly only two-thirds of Army and Marine Corps positions are open to women, nine out of ten positions—nearly all—are open to women in the Navy and Air Force. Thus, within the current system, women have a much greater chance of being promoted to top leadership positions in the Navy and Air Force than in the Army and Marine Corps (Harrell and Miller, 1997).

Figure 4.2
Distribution of Male Cadets' First-Choice Branch Preferences in 2007

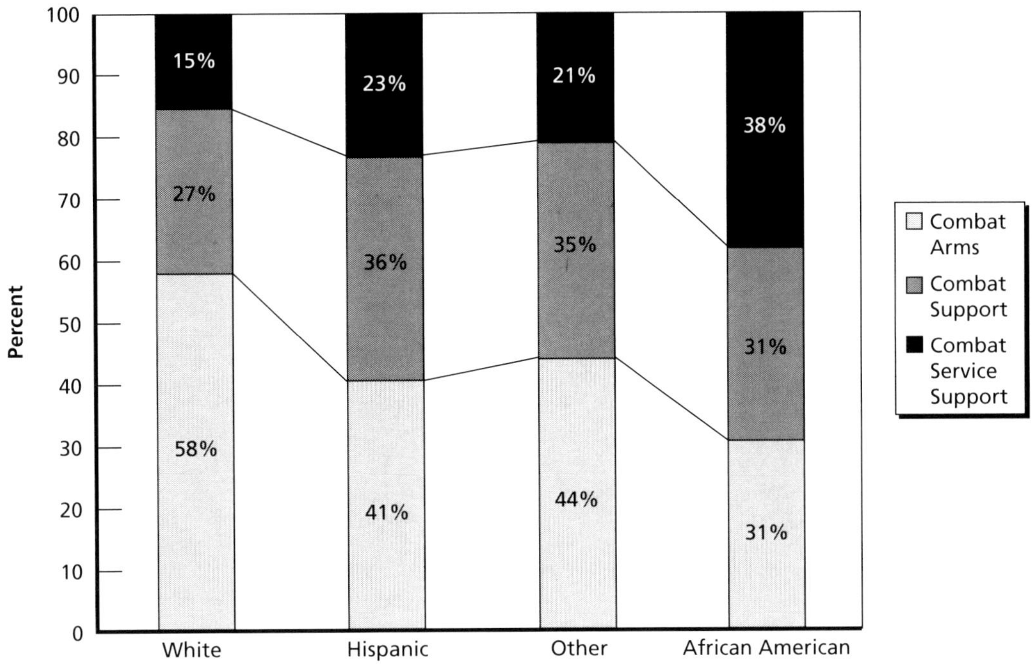

RAND *TR731-4.2*

Another factor that reinforces the career field divergence between majority and minority Army ROTC cadets is the fact that whites are much more likely than African Americans and Hispanics to prefer jobs within the same category (i.e., Combat Arms, Combat Support, etc.). Thirty-two percent of white cadets expressed career preferences within the same category in 2007, whereas only about 24 percent of Hispanic and African American cadets did so (see Table 4.4). More pointedly, while roughly 25 percent of white cadets picked Combat Arms branches as their top three career choices, only about 14 percent of Hispanic cadets and 6 percent of African American cadets followed this preference pattern. As a result, white cadets were more likely to be assigned to Combat Arms even if they did not receive their top career choice, while Hispanic and African American cadets were less likely to be assigned to Combat Arms because they spread their preferences across different branches.

Minority Cadets Tend to Rank Lower on Order of Merit Score

As explained above, the other major factor affecting career field selection among Army ROTC cadets (aside from individual job preferences) is the placement of cadets on the OML. In order to investigate the importance of this factor on selection outcomes, we analyzed the qualifications of cadets entering active duty service in 2007 by racial/ethnic group. We used the OMS, by which the Army ranks cadets on the OML, as the major qualification measurement. In general, white cadets earned higher scores than members of other racial/ethnic groups and thus tended to rank higher. The mean OML ranking for a white cadet

Figure 4.3
Distribution of Male Cadets' First-Choice Career Fields in 2007

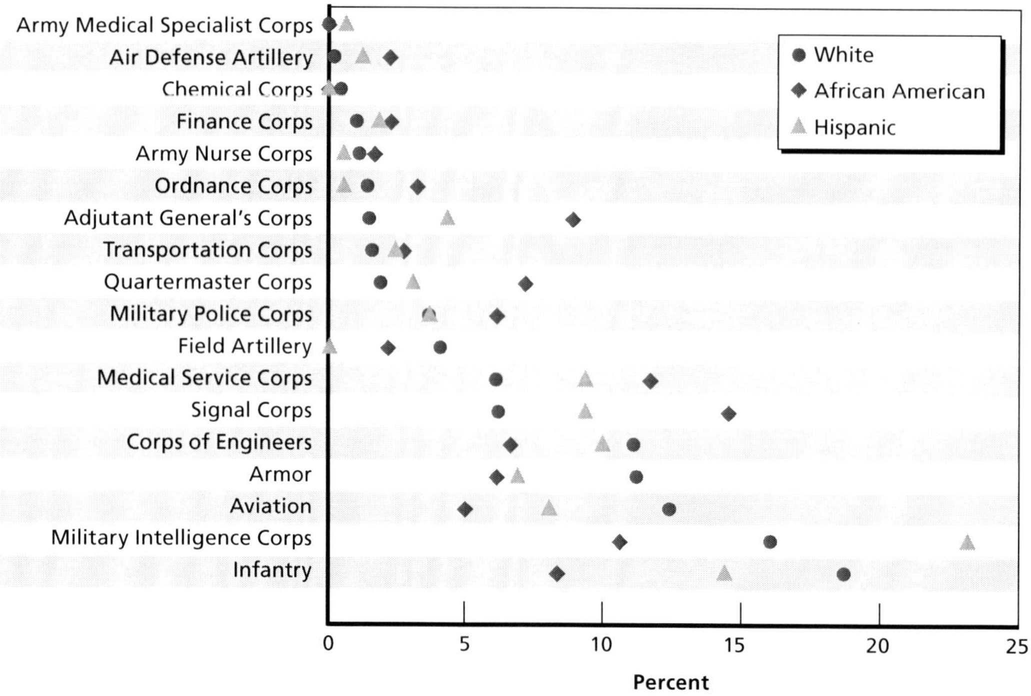

NOTE: The "Other" category is not included because its pattern closely resembles that of white cadets.
RAND *TR731-4.3*

Table 4.4
Category Preference Consistency Across Top Three Selections in 2007 (%)

Racial/Ethnic Group	Top Three Preferences in Same Category	Top Three Preferences in Combat Arms
White	32.16	24.62
Hispanic	24.38	13.75
Other	28.22	17.79
African American	24.58	6.15
Total	30.75	21.94

in 2007 was 1645, while the mean ranking was 2325 for an African American, 1917 for a Hispanic, and 2059 for other race/ethnicity cadets.[2] Additionally, while more than 10 percent of white cadets ranked in the top 10 percent of the OML, fewer than 3 percent of African Americans ranked as high.

The OMS distributions by race/ethnicity are shown in Figure 4.4. OMS values ranged from 60.7 to 98.4 in our dataset. The median OMS (the horizontal line in the middle of the

---

[2]   There were 3,803 total ranked Army cadets in the data.

respective box) of white cadets was nearly five points higher than the median OMS of African American cadets.

Most Cadets Received Their Top Preferences

Having established significant differences in the job preferences and qualifications of majority and minority ROTC cadets, we now examine the likelihood that cadets within different demographic groups receive their preferred career fields.

Overall, 54 percent of cadets received their first career field choice, 79 percent received one of their top three choices, and 86 percent received one of their top five choices (Table 4.5). Only 14 percent of the ROTC cadets in 2007 received a career field they did not ask for. These numbers do vary with race/ethnicity. A higher proportion of African American cadets received their top choice than any other racial/ethnic group: 63 percent compared with 55 percent of white cadets, 47 percent of Hispanic cadets, and 46 percent of cadets from other racial/ethnic groups. All racial/ethnic groups received one of their top three choices at roughly the same rate, though the Other race/ethnicity group was about 8 percent lower than the other three groups. The same is true for those receiving one of their top five choices, although again the Other race/ethnicity group is slightly lower.

This analysis indicates that most Army ROTC cadets, irrespective of their race/ethnicity and OMS, received one of their top five career field choices in 2007, and many received their first choice. Furthermore, African American cadets were the most likely among all racial/ethnic groups to receive their top choice.

Figure 4.4
Order of Merit Scores by Race/Ethnicity in 2007

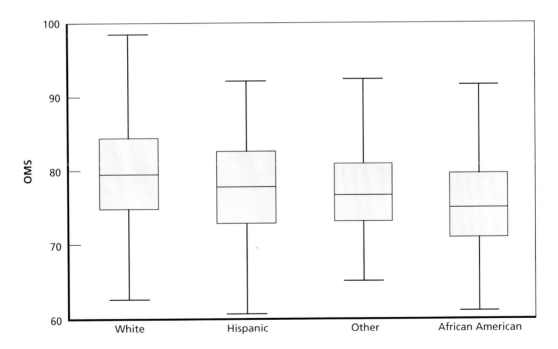

Table 4.5
Fraction of Career Field Preferences Matched by Race/Ethnicity in 2007

Race/Ethnicity	First Preference Matched	Top Three Preferences Matched	Top Five Preferences Matched
White	0.55	0.79	0.87
	(0.50)	(0.41)	(0.34)
Hispanic	0.47	0.80	0.86
	(0.50)	(0.40)	(0.35)
Other	0.46	0.71	0.83
	(0.50)	(0.45)	(0.38)
African American	0.63	0.80	0.87
	(0.49)	(0.40)	(0.34)
Total	0.54	0.79	0.86
	(0.50)	(0.41)	(0.34)

NOTE: Standard deviations in parentheses.

Though we have identified that minorities do not tend to prefer Combat Arms branches and that they are just as likely as white cadets to get their preferred branch, it is still possible that minorities might avoid Combat Arms branches because of competition from white cadets. To address this question, we examined the competitiveness of the branches selected by whites, Hispanics, African Americans, and other minority groups in 2007. One way to measure competitiveness is to compare the number of applications to the number of slots. Branches that have a small slot-to-applicant ratio would then appear to be quite competitive. When we do this, we find that Finance Corps was the most competitive branch and Chemical Corps was the least competitive branch. Not one of the four most-competitive branches in the 2007 Army ROTC selection process (Finance, Military Intelligence, Signal Corps, and Adjutant General's Corps) belonged to Combat Arms. Therefore, even though racial and ethnic differences in career preferences exist, minority cadets do not seem to be in danger of being crowded out of Combat Arms career fields because of competition from white cadets.

Measuring competitiveness in this way, however, does not tell the whole story. We have previously shown that minorities tend to have lower OML rankings and that they also tend to prefer Combat Support and Combat Service Support branches. If lower-ranking ROTC cadets anticipate competition for Combat Arms branches, they might strategically avoid those branches, causing Combat Arms branches to appear less competitive. This possibility is difficult to test in our data, but one way to examine the issue would be to look at how Combat Arms preferences vary with competitiveness (as measured by OMS). If minority cadets strategically avoid Combat Arms due to competitiveness, then minorities at the top of the OMS distribution should be more likely to choose Combat Arms branches than minorities at the bottom of the distribution. Figure 4.5 shows how the percentage of each group that selects a Combat Arms branch for their top choice varies with their place in the OMS distribution.

Figure 4.5
First-Choice Combat Arms Percentage Versus OMS Percentile in 2007

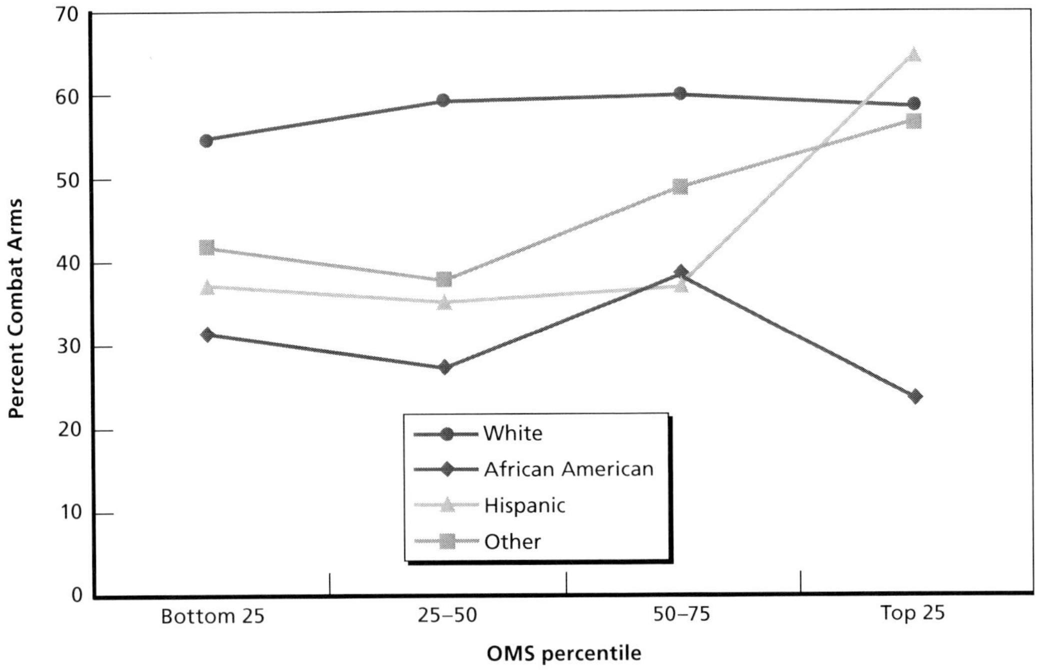

RAND *TR731-4.5*

In general, Hispanics and Other minority groups are more likely to select a Combat Arms branch as their OMS standing increases. For African American and white cadets, however, there is no clear relationship between OMS percentile and the percentage of cadets that select a Combat Arms branch for their top choice. African Americans throughout the OMS distribution are less likely than similarly competitive white cadets to opt for Combat Arms. This evidence indicates that racial/ethnic differences in branch choice are not strategic for African Americans, but rather reflect an actual divergence in preferences. Still, there were too few minorities in the dataset to arrive at any strong conclusions about the relationship between OMS and race-specific branch preferences.

Summary

The preceding analysis indicates that several factors are affecting minority representation in the senior ranks of the Army. One factor is cadets' branch preferences. Minority cadets generally prefer the Combat Support and Combat Support Service branches. Conversely, white cadets are much more likely than other races and ethnicities to list Combat Arms career fields as all of their top three choices, thus increasing their odds of being admitted to a Combat Arms branch.

Although there do tend to be racial/ethnic differences in branch preference, most cadets receive one of their top preferred branches regardless of their racial or ethnic group. Of the four racial/ethnic groups we analyzed, African Americans most often receive their top branch choice.

Finally, while it is possible that minorities perceive the Combat Arms braches as being more competitive than the support branches, our analysis has shown that this is not necessarily true. In addition, even highly competitive African Americans are less likely than whites to prefer Combat Arms branches, suggesting that minorities are not being crowded out of Combat Arms branches because of competition from white cadets.

Chapter Five will present the policy discussion and recommendations of our analysis.

Policy Discussion and Recommendations

A Comprehensive Study of the Classification System Is Needed

In this exploratory study, we have demonstrated that it is critical for the Army to increase minority representation in key career fields to improve the racial and ethnic diversity of its top military officers. But we also contend that there is a strong need for a more in-depth analysis of the Army branching process.

If, as our study suggests, minorities are indeed self-selecting into career fields with relatively limited promotion opportunities, why are they doing so? On the one hand, minority cadets could truly prefer different career fields than white cadets. In this case, policy should focus on ways to make combat career fields more appealing to minorities. On the other hand, minorities may not really prefer support career fields but rather may reason that they lack the OML ranking to get a more competitive career field (or they forecast a low probability of success in that career field). In this case, minority cadets might desire a Combat Arms career field but may opt for their most-preferred Combat Support or Combat Service Support career field thinking that they would never get a top Combat Arms assignment. This theory fits with both the even level of competitiveness across preferred career fields and the fact that African American cadets most often receive their top preference. However, the lack of a relationship between OMS and preference for a Combat Arms branch casts some doubt on the theory and makes a true difference in preferences more likely.

These two possibilities would have profoundly different policy prescriptions. Applying the wrong type of policy would likely do nothing to improve the diversity situation. Therefore, a comprehensive study is necessary to begin to understand the rationale behind minority choices. The Army will only be able to design appropriate policies when it understands which characteristics of Combat Arms branches appeal to whites and which characteristics of Combat Support and Combat Service Support branches appeal to minorities. Without this information, a policy that intends to mitigate group differences in branch preferences could easily make such differences worse. A more complete understanding of the determinants of individual career field decisions would enable the Army to offer incentives that appeal to minorities, and potentially improve long-run career outcomes.

Therefore, we recommend that DoD conduct a comprehensive study of the classification systems within all the services and commission sources. While the findings of our Army ROTC case study may be valid, the factors that explain low numbers of minorities within the senior officer ranks may differ for other services and commission sources.

Conclusions

In the case of the Army ROTC, the key career fields that lead to the highest positions in the service have disproportionately low numbers of minorities, and this lack of minority personnel stems primarily from the different selection preferences of minority and white cadets. While minority cadets generally rank lower on the OML than their white counterparts, this lower ranking does not keep minorities from getting their preferred career field. In fact, African American cadets received their top choice at a higher rate than white cadets did.

Given this state of affairs, providing equal opportunity with respect to career or promotion opportunities may not be enough to create a diverse senior military leadership. In order to achieve diversity at the highest levels, the Army has three options:

1. Select more generals from Combat Support and Combat Service Support occupations.
2. Select minority Combat Arms officers at a disproportionate rate.
3. Increase the number of minorities in Combat Arms.

The first option may not be expedient, because mission requirements drive the high promotion rate of Combat Arms officers. Combat Arms officers have skills that make them fit for command, and altering this pattern may negatively impact the Army's ability to provide national security. But if it is possible to promote more officers from Combat Support and Combat Service Support occupations without adversely impacting the mission, then the first option could contribute to increased diversity.

The second option is neither feasible nor legal, and indeed would negatively impact the order and discipline of the army. Prior research has shown that the desire to avoid compromising standards in the face of combat requirements transcends racial lines (Harrell et al., 1999; Hosek et al., 2001).

Finally, if the Army wishes to increase the level of diversity in the highest ranks, it needs to get more minorities to opt for Combat Arms. It is not immediately clear from this analysis how the Army can affect minority cadet career field decisions. The current bundle of incentives associated with Combat Arms seems to appeal disproportionately to whites. To increase the number of minorities who choose Combat Arms, the Army must find incentives that appeal to minorities and then tie those incentives to Combat Arms branches.

Currently, the Order of Merit system does not appear to prevent minorities from obtaining their preferred branches. If further research was able to highlight a policy option that could increase the number of minorities who prefer Combat Arms branches, such a policy could create an additional challenge—OML competitiveness could become an issue (since white cadets, who tend to rank higher, also tend to prefer Combat Arms). In that case, the Army would need to modify the Order of Merit system to see any gains in diversity.

Of course, policies aimed at changing outcomes of the OMS determination method would require a separate analysis, which is beyond the scope of this report. In this sense, any recommendation of this sort would be premature. A number of options are available to policymakers, but there is very little information on the most effective course of action. Army policymakers must understand more about the interaction between preferences and OML ranking in order to affect the racial/ethnic distribution of key career fields.

Finally, it is important to remember that career field assignment is just the first of many mileposts on the long road to becoming a general officer. The racial/ethnic makeup of any senior leadership will depend not only on career field choice, but future assignments, promotions, attrition, and numerous other factors. Hypothetically, if the Army forced every minority cadet into a Combat Arms branch, there would likely be great attrition (because they would be unsatisfied with such an outcome) and no gains in diversity. There is no guarantee that making Combat Arms more attractive to minorities will lead to large gains in diversity 30 years from now.

Still, if policymakers had a complete understanding of the determinants of cadet preferences for different commissioning services, researchers could estimate the number of cadets from each minority group necessary to reach the desired level of minority candidates for the most senior ranks. At this point, such analysis is premature because so little is known about cadet preferences for different career fields. In the future, a model of officer career flows in light of updated incentives will be important.

There are no easy solutions to the Army's continuing diversity challenges at the highest levels of command. All options involve trade-offs that the Army (and other services facing similar problems) will need to weigh. The Army may value fairness to the individual, while realizing that a perfectly merit-based system will have negative consequences down the road. The role of policy, then, is to be shrewd in balancing these competing needs to provide for the most robust national defense possible.

Detailed Description of Classification Processes

U.S. military classification processes vary significantly by service as well as by commission source. Most of these processes are described in detail below. With the exception of the Marine Corps, classification is separate for each source (i.e., there is no cross-source ranking; a number of slots are allocated to each source prior to classification, and the source then fills the slots).

Army

ROTC

In the fall of each year, cadets submit their branch choices in order of preference. Cadets can select branches they prefer with the exception of special branches: doctors, lawyers, and chaplains. Cadets must be medically qualified to select Combat Arms branches. At this time, cadets also have the option to volunteer for an additional three-year Active Duty Service Obligation (ADSO), after the three-year ROTC obligation, under the Branch, Post, or Graduate School for Service programs. Cadets indicate whether they are willing to extend their ADSO in order to receive their first or second branch preferences, Posting of Choice (Post), or graduate school. Cadets may request to participate in all three programs; however, cadets may only be selected for up to two of the programs.

Also in the fall, a cadet Order of Merit List (OML) is published using the Order of Merit scores (OMS). This OML produces the National OML Cadet rank 1-N. The OMS are based on three components, for a maximum possible total of 100 points: academic grade point average (40 possible points), leadership (45 possible points), and physical (15 possible points). Additional bonus points are available for specific accomplishments—for example, Warrior Forge Platoon Top Five equals 1 point added to the final OMS, and Recondo equals 0.5 point added to the final OMS. Once the national OML is released, cadets can adjust branch preferences.

The academic component is the cadet's cumulative GPA, including the ROTC GPA. The leadership component is made up of the following subcomponents and associated possible points:

- (6.75) Warrior Forge (WF) performance (leadership positions, attributes/skills/actions)
- (11.25) WF Platoon Tactics evaluation
- (4.5) WF land navigation (1st score)
- Professor of Military Science (PMS) experienced-based observations
- (6.75) PMS Military Science III cadet evaluation report OML
- (4.5) PMS accessions OML

- (4.5) PMS accessions potential comments
- (4.5) cadet training/extracurricular activities
- (2.25) language and cultural awareness.

The physical component is made up of the following subcomponents and associated possible points:

- (12.75) Army Physical Fitness Test (APFT)
- (0.75) swimming
- (1.5) varsity, intramural, or community team athletics.

After the cadets' branch preferences are submitted and the cadet OML ranking is determined, the Active Duty Branching Board determines cadet branches. Cadets on the active duty OML are branched in OML order, highest to lowest, their first three branch choices. If they fail to secure one of their top three choices, they receive their branch by the Department of the Army Branching Model (DABM)and the branching board.

Cadets can receive one of their top three choices by one of three methods: (1) The top 10 percent of the active duty OML receive their first branch choice, if eligible. Aviation candidates must have an approved flight physical and valid Alternate Flight Aptitude Selection Test (AFAST) score. The choices for all others are considered in order. (2) Cadets are either awarded their branch without needing to commit to a Branch ADSO or (3) use a Branch ADSO to secure their branch. If the desired branch has not yet filled to 50 percent of ROTC's allocation for that branch, cadets receive that branch without requiring a Branch ADSO. If the desired branch has been filled to 50 percent but not yet 100 percent, the cadet can be awarded the branch if he/she signed a Branch ADSO contract for that branch. One additional requirement is that no more than 65 percent of the branch's allocation can come from the top half of the active duty OML. Therefore, once the branch has been filled to 65 percent, the next cadet to receive that branch must be from the lower half of the OML and is required to accept Branch ADSO. For those cadets who fail to secure one of their top three branch choices, the DABM allocates an initial branch assignment to one of the branches that failed to fill to 100 percent by the first three methods. The DABM seeks to meet Army goals for quality distribution and other factors. It does consider cadet preference but will not assign one of their top three choices. A branching board member reviews each cadet assigned to his or her branch by the DABM. If the board member rejects a cadet, that cadet will be replaced by one rejected by a different branch. All ROTC cadets receive a basic branch assignment or are referred by the board back to the U.S. Army Cadet Command for commission review, to determine if the cadet should actually be commissioned.

After each cadet has been assigned a basic branch, the branching board determines which cadets will be required to be branch-detailed. Branch detailing is a process by which officers from some donor basic branches (Military Intelligence, Signal Corps, Finance Corps, Air Defense Artillery, and Adjutant General Corps) are required to serve their first three years in a recipient branch (Infantry, Armor, Field Artillery, or Chemical Corps). The board members from the donor branches identify the cadets to be detailed into each required recipient branch. Cadets can volunteer to be detailed into certain branches, and that factor is considered by the board. If nonvolunteers who received their branch with ADSO are detailed, the Branch ADSO will not be charged. The recipient branch board member either accepts the detailed cadet or the board member's branch accepts fewer detailed officers.

The final determination at the branching process is which future officers will receive their requested ADSO incentives for Posting of Choice (Post ADSO) and fully funded graduate school (Grad ADSO). For Grad ADSO, requests are awarded in the following order: Distinguished Military Graduates (DMGs); four-year scholarship cadets by OML; three-year scholarship cadets by OML; two-year scholarship cadets by OML; and nonscholarship cadets by OML. Only 20 percent of the aviation branch can receive Grad ADSO by OML, and Medical Service Corps cadets cannot receive Grad ADSO. Cadets who can receive Post ADSO are determined by Human Resources Command (HRC), based on Army requirements at the requested posts. When there are more requests than requirements, cadets are considered by OML, unless there is a timing requirement (a lower-listed cadet is available/required before a higher-listed cadet). Post ADSO requests can be made after branch results are published. Members of the Aviation branch cannot participate, because their posting will be based on aircraft.

U.S. Military Academy (USMA)

Like Army ROTC cadets, USMA cadets provide rank-ordered branch preferences and also can volunteer for the Branch for Service program to extend their ADSO in order to receive their branch preference. The first 75 percent of branch allocations will be assigned according to the cadet performance rank (CPR) branching process. This percentage may be adjusted based on the needs of the Army. The CPR is the rank ordering of cadet performance scores (CPS), which are made up of an academic program score (APS), a military program score (MPS), and a physical program score (PPS). The APS makes up 55 percent of the CPS and is based on cadets' cumulative GPA, excluding core military science and physical education courses. The MPS makes up 30 percent and is calculated by averaging grades in summer training, military duty performance, and military science courses. The PPS makes up 15 percent and includes grades from physical education courses, fitness test scores, and competitive athletic performance.

After the 75 percent target is attained, branch allocations are assigned to Branch for Service volunteers in CPR order. When no further volunteers remain, the remaining branch allocations are assigned to cadets in CPR order. One of the goals of the branching process is to ensure that approximately 80 percent of male cadets and 20 percent of female cadets are assigned to Combat Arms.

Air Force

Unlike the Army, the Air Force is divided into line and nonline officers. Nonline officers consist of the Medical, Nurse, Biomedical Services, Medical Service, and Chaplain corps and are classified through selection board processes. Line officers are divided into two categories: rated and nonrated. Rated officers consist of pilots, combat systems officers (formerly known as navigators), and air battle managers. Cadets interested in becoming rated officers meet a selection board at their respective source of commission and are classified based on that commissioning source's criteria and AF requirements. The classification process for nonrated line officers (except for judge advocates, combat rescue officers, and special investigators) is described below.

AFROTC

AFROTC detachment commanders submit an AF Form 53 for all cadets detailing their qualifications—including academic major, coursework, foreign language skills, and GPA—and the cadets' desired Air Force Specialty Code (AFSC) (cadets provide six in rank order). HQ AFROTC determines a national order or merit (OM) for all cadets based on their GPA, field training scores, physical fitness test scores, SAT equivalency score, and detachment ranking. The OM and information from the AF Form 53 are imported into a computer model for classification.

The computer model factors in Air Force requirements for each AFSC having specified degree requirements; others have specific course requirements while others may have only desired educational requirements. A computer algorithm maximizes the matches to AFSC targets while factoring in cadet qualifications, OM, Air Force requirements, cadet preferences, and special qualifications (e.g., foreign languages, prior service, desire to join a spouse).

Each nonrated cadet then receives an AFSC.

U.S. Air Force Academy (USAFA)

The Board Order of Merit (BOM) plays a significant role in USAFA cadet classification. The AFSC General Board consists of two panels, each with five officers. Cadets are split between these two panels, with each panel reviewing records for half the graduating USAFA class. Panel members rate each cadet on a scale of 6 to 10, based on the cadet's academic, military, and athletic performance, as well as leadership potential. The individual panel member scores are summed for an overall panel score. After scoring is completed, each panel standardizes the scores and the two panels then merge their results to create a rank-ordered list of all cadets. The BOM rating and other information, such as academic major, coursework, foreign language skills, GPA, and the cadets' desired AFSC (cadets provide six preferences in rank order) are gathered on each cadet who desires classification into a nonrated AFSC. This information is imported into a computer model (similar to the one used to classify AFROTC cadets). The model considers Air Force requirements, cadet qualifications, BOM, cadet preferences, and special qualifications (e.g., foreign languages, prior service).

Navy

NROTC

The NROTC program begins with three overarching options: Navy, Nurse Corps, and Marine Corps. Midshipmen enrolled in the Nurse Corps option are assigned their active-duty community, Navy Nurse Corps, when they enroll in the NROTC program. A few midshipmen in the Marine Corps option are designated as aviators while in the NROTC program. However, the majority of NROTC midshipmen in the Marine Corps option receive their career field (their Military Occupational Specialty or MOS) at the completion of The (Officer) Basic School (TBS). For Navy option midshipmen, a separate career field assignment process takes place during their senior year. The following describes the assignment system of those who choose to go into the Navy.

The NROTC Program assigns most Navy option midshipmen to the surface, submarine, or (naval) aviation career field designators. These career fields, in addition to special warfare, constitute the Navy's unrestricted line (URL) community. Navy option midshipmen only enter the restricted line or staff corps (RL/SC) communities if they are not physically qualified for the URL.

During August of each year, all NROTC units are required to submit service assignment packages for all Navy option midshipmen scheduled to graduate in the next fiscal year. Those packages include the students' desires (to include ranking of choices for each of the major warfare communities—surface, submarine, and aviation), information about student qualifications and performance, and Professor of Naval Science (PNS) ranking and recommendations. Students provide six choices, with the first five listings being URL communities and the last choice being a RL/SC community in case of medical disqualification.

Ranking scores are computed, and the Naval Service Training Command (NSTC) assigns all students a corresponding score based on their performance and PNS points and recommendations. A panel of representatives from the surface, submarine, and aviation communities ensures equality during the selection process. Additionally, a special review is conducted by community representatives for those desiring to enter the Special Warfare/SEAL or Special Operations/Explosive Ordinance Disposal communities. Selections to the submarine and surface nuclear communities require additional review and approval by processes external to NROTC.

The NROTC service assignment panel meets, rank-orders individuals, assigns an order of merit, and assigns each midshipman a URL designator. This process takes into consideration the candidate's desire, the needs of the Navy, and order of merit. Once the top 25 percent have been assigned, the remaining individuals and their choices are reviewed to ensure that the goal of every community goal has been achieved. Upon completion of the assignment process, the Director of Officer Development and Director of Officer Development Military Operations, as an executive review panel, conduct a final quality control and verification. Once executive review is completed, the results are presented to Commander, Naval Service Command (CNSTC) for final approval. All changes and amendments to the list are also approved by CNSTC.

U.S. Naval Academy (USNA)

In the fall, USNA midshipmen provide six rank-ordered community preferences for which they are eligible and have been medically screened. Special qualifications required for certain communities include:

- Navy pilot/naval flight officer (NFO): medically qualified for aviation; Aviation Selection Test Battery (ASTB)
- Surface warfare officer: medically qualified URL
- Submarines: medically qualified submarines; screened by Naval Reactors for program interview (Naval Reactors is the U.S. government office that has comprehensive responsibility for the continued safe and reliable operation of the United States Navy's nuclear propulsion program)
- Special warfare (SEAL)/explosive ordnance disposal (EOD): medically qualified; completion of physical screening test
- U.S. Marine Corps: medically qualified URL; color-deficient midshipmen are eligible for USMC ground assignment.

These midshipmen are then screened by locally run community assignment boards. These boards assess midshipmen based upon their qualifications, USNA academic and military performance, and documented aptitude. Midshipmen's records are initially screened by the board representing the community most preferred by the midshipman, and initial selection recommendations are made. Midshipmen not selected for their first-choice community are then considered by their second-choice community. This process continues until all midshipmen have been assigned to a community for which they are qualified.

A review board (chaired by the USNA commandant) then reviews the community assignment recommendations, including the distribution of midshipmen preferences, demographics, and performance ranking. If necessary, the review board will determine best-fit midshipmen reassignments to ensure that Chief of Naval Operations–directed minimum accessions requirements of all communities are satisfied. The review board then adjourns to permit community assignment boards to select remaining accessions until CNO-directed goals are met. The review board shall reconvene as required for review or deliberation prior to forwarding service assignment recommendations to the Superintendent for approval.

Marine Corps

Unlike in the other services, newly commissioned officers in the Marine Corps attend TBS for six months before the branching process begins. At the conclusion of TBS, Marine Corps officers submit career field preferences, and candidates are given a score that determines their OML ranking. As in the other services, scores are composed of academic grades, military education grades, and leadership evaluations based on field exercise performance. However, each of these components is derived solely from the candidate's time at TBS.

To ensure quality personnel in all career fields, the Marine Corps uses the Rule of Thirds: for each career field, one-third of the allotments must come from the top third of the OML, one-third must come from the middle third, and the remaining one-third must come from the bottom third.

Officer Training School/Officer Candidate School (OTS/OCS)

Branching processes can also differ when an officer's source of commission is OTS/OCS. This is not the case for the Marine Corps, because all newly commissioned Marine Corps officers attend TBS. However, the Army OCS uses quarterly boards to review candidates and assign branches. Unlike other commissioning sources, the Army OCS takes into account Army experience, in addition to such factors as candidate preference and the needs of the Army. The Air Force OTS process is discussed in detail below. We did not explore the details of the Navy OCS branching process.

Air Force OTS

Civilians; active duty enlisted Air Force members; and enlisted members of the Reserves, National Guard, and other services may apply for Air Force Basic Officer Training School provided they meet the requirements for commissioning. Applications are considered by a selection board consisting of several panels of officers at the grade of colonel.

Selection boards are conducted by the Air Force Recruiting Service, with multiple boards held each year. Individuals meet panels based on the AFSC for which they apply. Rated candidates are considered only by the rated board and compete only with other individuals applying for rated positions. Once the board is complete, candidates are classified as pilots, combat systems officers, or air battle managers. Nonrated candidates are considered for technical or nontechnical boards, depending on their academic majors. Each board establishes a board OML based on board score. Once the nonrated boards are complete, individuals selected to receive commissions are classified manually using qualifications for each AFSC as outlined in the Air Force Officer Classification Directory. Classification is completed using the board score, academic qualifications, individual desires, and the Air Force's requirements for each AFSC.

Bibliography

Aguilera, M. B., "The Impact of Social Capital on Labor Force Participation: Evidence from the 2000 Social Capital Benchmark Survey," *Social Science Quarterly,* Vol. 83, No. 3, 2002, pp. 853–874.

Akerlof, George A., "Social Distance and Social Decisions," *Econometrica,* Vol. 65, 1997, pp. 1005–1028.

Baldwin, J. Norman, "Female Promotions in Male-Dominant Organizations: The Case of the United States Military," *The Journal of Politics*, Vol. 58, No. 4, 1996a, pp. 1184–1197.

————, "The Promotion Record of the United States Army: Glass Ceilings in the Officer Corps," *Public Administration Review*, Vol. 56, No. 2, March–April 1996b, pp. 199–206.

Baldwin, J. Norman, and Bruce A. Rothwell, "Glass Ceilings in the Military," *Review of Public Personnel Administration*, Vol. 13, No. 5, 1993, pp. 5–26.

Becker, G. S., "Investment in Human Capital: A Theoretical Analysis," *The Journal of Political Economy,* Vol. 70, No. S5, 1962, p. 9.

Becton, Julius W. Jr., et al., Amici Curiae in Support of Respt. At 1, *Gratz v. Bollinger*, 123 S. Ct. 2411 (2003) and *Grutter vs. Bollinger*, 123 S. Ct. 2325 (2003).

Bertrand, Marianne, Erzo F. P. Luttmer, and Sendhil Mullainathan, "Network Effects and Welfare Cultures," *Quarterly Journal of Economics,* Vol. 115, No. 3, 2000, pp. 1019–1055.

Bielby, William T., "The Structure and Process of Sex Segregation," in Richard R. Cornwall and Phanindra V. Wunnava (eds.), *New Approaches to Economic and Social Analyses of Discrimination*, Portsmouth, N.H.: Praeger Greenwood, 1991, pp. 97–112.

Blau, Peter M., and Otis Dudley Duncan, *The American Occupational Structure*, New York: The Free Press, 1967.

Blau, Peter M., John W. Gustad, Richard Jessor, Herbert S. Parnes, and Richard C. Wilcock, "Occupational Choice: A Conceptual Framework," *Industrial and Labor Relations Review*, Vol. 9, No. 4, July 1956, pp. 531–543.

Borjas, G. J., "Ethnicity, Neighborhoods, and Human-Capital Externalities," *American Economic Review*, Vol. 85, 1995, pp. 365–365.

Chestang, Carlen J. Jr., *The U.S. Army Officer Corps: Changing with the Times*, Carlisle Barracks, Pa.: U.S. Army War College Strategy Research Project, March 15, 2006.

Coffey, David, *The Encyclopedia of the Vietnam War: A Political, Social, and Military History*, Spencer C. Tucker (ed.), Oxford, UK: ABC-CLIO, 1998.

Defense Human Resources Board, "Senior Leadership Diversity Forecast," internal briefing, Department of Defense, December 2005.

Defense Manpower Data Center, PERSTEMPO database, 2006.

Denning, Daniel B., Acting Assistant Secretary of the Army (Manpower and Reserve Affairs), "Graduate School, Branch, or Post of Choice for Service Programs," Memorandum for Commander, U.S. Army Cadet Command, Fort Monroe, Va., Sept. 14, 2006.

Diversity Working Group, internal presentation at Diversity Working Group meeting, July 2005.

DMDC—*See* Defense Manpower Data Center.

Groner J., "In 'Grutter v. Bollinger' Amicus Avalanche, One Brief Stood Out," *Legal Times*, July 2, 2003.

Harrell, Margaret C., Megan K. Beckett, Chiaying Sandy Chien, et al., *The Status of Gender Integration in the Military: Analysis of Selected Occupations*, Santa Monica, Calif.: RAND Corporation, MR-1380-OSD, 2002. As of July 21, 2009:
http://www.rand.org/pubs/monograph_reports/MR1380/

Harrell, Margaret C., Shelia Nataraj Kirby, Jennifer S. Sloan, et al., *Barriers to Minority Participation in Special Operations Forces*, Santa Monica, Calif.: RAND Corporation, MR-1042-SOCOM, 1999. As of July 21, 2009:
http://www.rand.org/pubs/monograph_reports/MR1042/

Harrell, Margaret C., and Laura L. Miller, *New Opportunities for Military Women: Effects upon Readiness, Cohesion, and Morale,* Santa Monica, Calif.: RAND Corporation, MR-896-OSD, 1997. As of July 21, 2009:
http://www.rand.org/pubs/monograph_reports/MR896/

Hosek, James, and Christine E. Peterson, *Serving Her Country: An Analysis of Women's Enlistment,* Santa Monica, Calif.: RAND Corporation, R-3853-FMP, 1990. As of July 21, 2009:
http://www.rand.org/pubs/reports/R3853/

Hosek, Susan D., Peter Tiemeyer, M. Rebecca Kilburn, Debra A. Strong, Selika Ducksworth, and Reginald Ray, *Minority and Gender Differences in Officer Career Progression*, Santa Monica, Calif.: RAND Corporation, MR-1184-OSD, 2001. As of July 21, 2009:
http://www.rand.org/pubs/monograph_reports/MR1184/

Kalleberg, A. L., and A. B. Sørensen, "The Sociology of Labor Markets," *Annual Review of Sociology,* Vol. 5, No. 1, 1979, pp. 351–379.

Kilburn, M. Rebecca, and Jacob Alex Klerman, *Enlistment Decisions in the 1990s: Evidence from Individual-Level Data,* Santa Monica, Calif.: RAND Corporation, MR-944-OSD/A, 1999. As of July 21, 2009:
http://www.rand.org/pubs/monograph_reports/MR944/

Kilduff, Martin, "The Friendship Network As a Decision-Making Resource: Dispositional Moderators of Social Influences on Organizational Choice," *Journal of Personality and Social Psychology,* Vol. 62, No. 1, 1992, pp. 168–180.

Kraus, Amanda, and Martha Farnsworth Riche, *Air Force Demographics: From Representation to Diversity*, Alexandria, Va.: CNA, CRM D0012197.A2, 2006.

Logan, John Allen, "Rational Choice and the TSL Model of Occupational Opportunity," *Rationality and Society*, Vol. 8, May 1996, pp. 207–215, 229–230.

Lubold G., "More Minority Officers Needed, Rumsfeld Says," *Air Force Times*, March 13, 2006.

Mani, A., and C. H. Mullin, "Choosing the Right Pond: Social Approval and Occupational Choice," *Journal of Labor Economics,* Vol. 22, No. 4, 2004, pp. 835–861.

Mason, LTC E. J., *Diversity: 2015 and the Afro-American Army Officer*, Carlisle, Pa., U.S. Army War College Strategy Research Project, 1998.

Meek, Kendrick B., "Strength in Diversity: Until Officer Corps Catches Up to Enlisted Ranks, Military Won't Be Truly Integrated," *Army Times*, November 19, 2007.

Moore B. L., and S. C. Webb, "Perceptions of Equal Opportunity Among Women and Minority Army Personnel," *Sociological Inquiry*, Vol. 70, No. 2, 2000, pp. 215–239.

Mosher, F. C. (ed.), *Democracy and the Public Service,* New York: Oxford University Press, 1982.

Moskos, Charles C., and John Sibley Bulter, *All That We Can Be: Black Leadership and Racial Integration the Army Way*, New York: Basic Books, 1996.

Reskin, Barbara F., "Labor Markets as Queues: A Structural Approach to Changing Occupational Sex Composition," in Joan Huber (ed.), *Micro-Macro Linkages in Sociology*, Newbury Park, Calif.: Sage Publications, 1991.

Rostker, Bernard D., *I Want You! The Evolution of the All-Volunteer Force*, Santa Monica, Calif.: RAND Corporation, MG-265-RC, 2006. As of August 21, 2009:
http://www.rand.org/pubs/monographs/MG265/

Sørensen, Aage B., and Arne L. Kalleberg, "An Outline of a Theory of the Matching of Persons to Jobs," in Ivar Berg (ed.), *Sociological Perspectives on Labor Markets*, New York: Academic Press, 1981.

Sewell, William H., Archibald O. Haller, and Alejandro Portes, "The Educational and Early Occupational Attainment Process," *American Sociological Review*, Vol. 34, February 1969, pp. 82–92.

Stewart, James B., and Juanita M. Firestone, "Looking for a Few Good Men: Predicting Patterns of Retention, Promotion, and Accession of Minority and Women Officers," *American Journal of Economics and Sociology*, Vol. 51, No. 4, October 1992, pp. 435–458.

Weber, Max, *The Theory of Social and Economic Organization,* A. M. Henderson and T. Parsons (eds.), Glencoe, Ill.: Free Press, 1947.

Wilson, Michael J., James B. Greenless, Tracey Hagerty, Cynthia V. Helba, D. Wayne Hintze, and Jerome D. Lehnus, *Youth Attitude Tracking Study: 1999 Propensity and Advertising Report*, Arlington Va.: Defense Manpower Data Center, 2000.